OLIVER TRACTORS

PHOTO ARCHIVE

OLIVER TRACTORS
PHOTO ARCHIVE

Photographs from the
Floyd County, Iowa Historical Society;
Bakken and Higgins Collections of the
Shields Library, University of California, Davis

Edited with introduction by
P. A. Letourneau

Iconografix
Photo Archive Series

Iconografix
PO Box 609
Osceola, Wisconsin 54020 USA

Library of Congress Card Number 93-61074
ISBN 1-882256-09-3

95 96 97 98 99 00 5 4 3 2

Cover and book design by Lou Gordon, Osceola, Wisconsin
Digital imaging by Pixelperfect, Madison, Wisconsin

Printed in the United States of America

Book trade distribution by Voyageur Press, Inc. (800) 888-9653

PREFACE

The histories of machines and mechanical gadgets are contained in the books, journals, correspondence and personal papers stored in libraries and archives throughout the world. Written in tens of languages, covering thousands of subjects, the stories are recorded in millions of words.

Words are powerful. Yet, the impact of a single image, a photograph or an illustration, often relates more than dozens of pages of text. Fortunately, many of the libraries and archives that house the words also preserve the images.

In the *Photo Archive Series*, Iconografix reproduces photographs and illustrations selected from public and private collections. The images are chosen to tell a story—to capture the character of their subject. Reproduced as found, they are accompanied by the captions made available by the archive.

The Iconografix *Photo Archive Series* is dedicated to young and old alike, the enthusiast, the collector and anyone who, like us, is fascinated by "things" mechanical.

ACKNOWLEDGMENTS

The photographs and illustrations appearing in this book have been selected from three collections: the Floyd County Historical Society, Charles City, Iowa; and the Bakken and Higgins Collections of the Shields Library, Special Collections, University of California, Davis. We appreciate the cooperation of both institutions.

Oliver Hart-Parr Row Crop. 1930.

INTRODUCTION

By 1929, the American farm equipment industry was dominated by International Harvester and Deere & Company, the principal full-line manufacturers. In order to compete with these giants, many venerable, smaller companies sought new partners. The Oliver Farm Equipment Company was formed in April 1929, through the merger of American Seeding Machine Company, Hart-Parr Tractor Company, Nichols & Shepard Company, and Oliver Chilled Plow Works. Each of the companies offered a line of machinery complementary to the others, and that gave Oliver a full range of equipment.

Hart-Parr Tractor Company was the first successful manufacturer of gasoline tractors. Production began in 1903 and, until 1911, Hart-Parr led the industry in sales. Following the merger that created Oliver, production of the Hart-Parr tractor line continued. A new line of "Oliver Hart-Parr" tractors emerged in mid-1930, led by the "Row Crop," Oliver's first general purpose tractor. Equipped with a 4-cylinder vertical engine, rated at 27 belt and 18 drawbar horsepower, the Row Crop featured a single front wheel. By 1932, dual front wheels, optional high-clearance, and adjustable front axles were introduced. In late 1930, Oliver also introduced two standard tractors: the Model 18-28, equipped with the same engine as the Row Crop; and the Model 28-44, equipped with a larger vertical 4-cylinder engine. The 18-28 and 28-44 were also offered in industrial, rice, and "Western Special" versions. The 18-28 and the Row Crop remained in production until 1937. The 28-44 was transformed into the Model 90, a distillate fuel tractor that remained in production until 1952, and the Model 99, a high-compression gasoline tractor that was built until 1957.

In 1935, Oliver introduced the Model 70. Built in standard, orchard, industrial, row-crop, and high-clearance row-crop configurations, it was offered with a choice of 6-cylinder engines: a high-compression gasoline version (HC); or a kerosene/distillate fuel version (KD). Fitted with the HC engine, the early Row Crop 70 was rated at maximum 26.57 brake and 18.03 drawbar horsepower. Oliver offered a 4-speed transmission as standard equipment, and an optional

6-speed transmission for the Standard and Orchard 70. In 1937, Oliver introduced the Model 80, a larger version of the Model 70. Available in standard and row-crop configurations, it was equipped with a choice of 4-cylinder HC or KD engines. The KD engine delivered maximum 36.33 brake and 27.13 drawbar horsepower. In 1940, Oliver offered the Diesel 80, equipped with a Buda-Lanova engine. In 1940, Oliver also introduced the Model 60, a smaller version of the Model 70. Available in standard, industrial, and row-crop configurations, it was equipped with a choice of 4-cylinder gas or distillate engines. Production of the 70 and 80 series tractors ended in 1948, with the introduction of the Oliver 77 and 88. The Oliver 60 was replaced by the Oliver 66, in 1949.

The Oliver 66, 77, and 88 were available in standard, orchard, industrial, row-crop, and high-clearance configurations. Power options for the Oliver 77 and 88 included: 6-cylinder HC and KD engines; LP-Gas; and, from 1949, an Oliver 6-cylinder diesel. Displacement varied, with the Oliver 88 offering larger and more powerful engines. The smaller Oliver 66 was equipped with a choice of 4-cylinder HC or KD engines or, beginning in 1951, a 4-cylinder diesel. A 6-speed transmission was standard on all three models. Production of the Oliver 66, 77, and 88 standard and industrial tractors ended in 1952, while production of row-crop tractors ended in 1954.

Between 1954 and 1958, Oliver offered the Super Series line of Super 44, 55, 66, 77, 88, and 99 tractors. The Super 55, Oliver's first utility tractor, was offered with a choice of 4-cylinder gas or diesel engines. The Super 66, 77, and 88 were built in standard, orchard, industrial, row-crop, and high-clearance configurations. The Super 66 was offered with either a 4-cylinder gas or diesel engine. Power options for the Super 77 and Super 88 included 6-cylinder gas, LP-Gas, and diesel engines. The Super 44 was a small utility tractor equipped with a 4-cylinder Continental gas engine. The Super 99 was adapted from the Standard 99, and was offered with a gas motor or choice of diesel engines: Oliver's 6-cylinder diesel or a 3-cylinder, 80 horsepower General Motors diesel with torque converter. The Super 44 and Super 99 were built in 1957 and 1958 only.

The Super Series gave way to the Oliver 440, 550, 660, 770, 880, 950, 990, and 995 tractors. Distinctly different in styling from the Super Series, they were the last Oliver tractors offered before the company was purchased by White Motor Corporation, in November 1960. The Oliver 440, built in 1960 only, and Oliver 550 were utility tractors. The 550, offered with gas or diesel 4-cylinder engines, saw duty in farm and industrial applications through the end of its production run in 1970. The Oliver 660, offered with narrow and adjustable front axles, and in 4-cylinder gas and diesel versions, remained in production until 1964. The 770 and 880 were built as standard, industrial, orchard, row-crop, and high-clearance tractors, available with 6-cylinder gas, LP-Gas, or diesel engines. The 770 remained in production until 1967, while 880 production ended in 1963. The Oliver 950, 990, and 995 were comparable standard and industrial tractors, but were equipped with different engines: the 950 featured an Oliver 6-cylinder diesel; the 990 and 995, General Motors 3-cylinder diesels. The 950 and 990 were equipped with standard transmissions, while the 995 featured a torque converter transmission that offered a speed range of 0.9 to 13.8 mph.

Oliver introduced the Models 1800 and 1900 in 1960, and followed with the Model 1600 in 1962. Built until 1964, this series offered standard, industrial, wheatland, riceland, row-crop, high-clearance, and four wheel-drive models. The Oliver 1600 and 1800 were offered with a choice of 6-cylinder gas, LP-Gas, and diesel engines. The Oliver 1900 was equipped with a General Motors 4-cylinder diesel. In 1964, Oliver updated the 1600, 1800 and 1900 tractors, boosting their horsepower output to create the Oliver 1650, 1850 and 1950. Offered in virtually the same configurations as their predecessors, these are the last Oliver wheel tractors included in this book.

In 1944, Oliver purchased the Cleveland Tractor Company, manufacturer of Cletrac crawler tractors. The Oliver name eventually supplanted that of Cletrac, and the company continued to build the "OC" line of crawler tractors until 1965. We include a selection of crawler photos among the Oliver industrial tractors.

In 1962, White Motor Company purchased Canada's Cockshutt Farm Equipment Company and, in 1969, purchased Minneapolis-Moline. For a period of time, certain Oliver models were marketed under the Cockshutt and Minneapolis-Moline names. The White name began to replace those of its subsidiaries, and by 1976 the proud name of Oliver had disappeared entirely.

Wide-tread Row Crop with integral cultivator. 1931.

Row Crop with front mounted furrow guide.

Row Crop with central mounted dozer blade.

Row Crop Model 18-27 with integral planter.

Oliver Hart-Parr Model 28-44 and 4-bottom Oliver plow.

OLIVER
HART-PARR
HART-PARR

Oliver Hart-Parr Model 18-28 powers baler, as Oliver Row Crop 70 moves hay with sweep rake.

Studio photograph of Row Crop 60 equipped with check-row corn planter.

Oliver Standard 60 and Oliver No. 100 2-bottom plow.

Row Crop 60 and integral 2-row cultivator.

Standard 60 and peg-tooth harrow.

Oliver Hart-Parr Orchard 70 tractor.

Row Crop 70 and 2-bottom plow.

Two views of a "4-Wheel" Row Crop 70, with high-clearance adjustable front axle. September 1938.

Orchard 70 and tandem disc harrow.

Orchard 70 pulling an Oliver Iron Age sprayer through a California orchard.

Oliver Hart-Parr Row Crop 70 with optional rubber tire single front wheel.

An experimental Row Crop 70 with rubber tires and cast disk wheels. (Note the two different grille patterns.)

Row Crop 70 tractor with Oliver 2-row potato digger. Idaho Falls, Idaho.

Row Crop 70 and Oliver Grain Master combine. 1938.

Oliver Row Crop 70 at Iowa State University.

Row Crop 70 pulling hydraulic scraper.

Row Crop 70 and No. 1495 beet and bean cultivator with side dressing attachment.

Two views of an experimental 70 half-track.

Two views of a Row Crop 80 with Buda-Lanova diesel engine.

OLIVER
OLIVER
80
ROW CROP

Two Row Crop 80s and Model 15 Grain Master combines.

Row Crop 80, with Buda-Lanova diesel engine, and adjustable front axle.

Fleetline Row Crop 66 with standard dual front tires. 1949.

Fleetline Standard 77.

Side and rear views of a Standard 77 with full orchard fenders and cowling.

Fleetline Row Crop 77 and Oliver Corn Master picker.

Row Crop 77 Diesel, with adjustable front end, and spring-tooth harrow.

Row Crop 88 Diesel with adjustable front end.

Fleetline Row Crop 88.

Row Crop 88 Diesel and Oliver corn picker.

Standard 88 Diesel and Oliver tandem disc.

Row Crop 88 Diesel supplying power to mill.

Orchard 88 Diesel in California orchard.

Super 55 utility tractor photographed at the factory.

Super 55 and chisel plow.

Super 55 and No. 330 tandem disc harrow.

Super 55 pulling a 2-row planter.

Oliver Super 66.

Super 77, with optional adjustable front axle, and deck mower.

Standard Diesel Super 77 with optional belt pulley. 1954.

High-clearance Super 88 Diesel.

Oliver 550 G (gas) tractor. 1962.

Model 550 G and mower. 1962.

OLIVER
OLIVER
OLIVER
550
IMPORTANT

OLIVER
550

Model 550 G with Model 1609 front end loader. 1962.

Orchard 770 Diesel.

Oliver 770 G with cotton picker and wagon.

OLIVER 770
OLIVER

Front and rear views of a high-clearance 770 G.

U.S. SAVINGS

Model 770 Gas with narrow front.

High-clearance Model 880 Diesel.

LP-Gas Model 880.

Oliver 995 Lugmatic Diesel featuring a General Motors 3-cylinder engine.

Model 1600 D (diesel) narrow front. 1963.

Standard axle 1600 D.

Oliver 1600 D hitched to Oliver 570 spreader, and Oliver 1600 G equipped with Model 1609 front end loader.

Oliver 1600 D, Model 1800 G, and Model 1900 D 4WD (four-wheel drive).

Rear view of a Model 1600. 1962.

Standard Model 1600 G. 1962.

Two views of a Model 1600 D 4WD. 1963.

DIESEL
HYDRA-POWER DRIVE
OLIVER
1600

Oliver 1800 G 4WD. 1962.

Model 1800 G with Oliver 96 grain drill.

Rear view of 1800 G with 6-bottom plow. 1962.

Standard Model 1800 D. 1962.

Potato diggers incorporating Model 1800 D tractor. 1962.

Interior and exterior views of a cab fitted to a Model 1800 G. 1962.

OLIVER
1800
OLIVER

Standard Model 1900 with optional canopy.

Front view of a Model 1900 D 4WD. 1963.

Two views of a Model 1900 D 4WD. 1962.

DIESEL
OLIVER
1900

Model 1650 D. 1964.

Model 1650 D pulling Oliver Model 620 baler.

Model 1650 G with adjustable front axle. 1964.

Model 1650 LP-Gas with adjustable front axle. 1964.

Model 1650 D, with optional cab, pulling Oliver 347 chisel plow.

Oliver 1650 G, with adjustable front axle, and plow.

Model 1650 G pulling Model 620 baler and wagon.

Standard Model 1850 LP-Gas with factory cab. 1964.

Two views of a Model 1850 G plowing. 1964.

OLIVER
1850

Model 1850 LP-Gas with adjustable front axle. 1964.

Close-up view of 1850 LP-Gas engine. 1964.

Standard Model 1850 D and semi-mounted 6-bottom plow. 1964.

Two views of a Model 1950 D 4WD.

1950
OLIVER

Oliver 1950 D 4WD and tandem disc.

OLIVER 1950

Oliver Hart-Parr 18-28 industrial tractor on hard rubber or "zero pressure" tires.

Model 900 Industrial.

OLIVER
66
INDUSTRIAL
OLIVER
66

Two views of an Oliver Super 99 Diesel with loader. 1958.

DIESEL

Two views of an Oliver OC-6 crawler tractor.

DIESEL OLIVER POWER
OLIVER OC-6
OLIVER

Model OC-6 skidding logs.

Model OC-6 picking corn.

Model OC-4 dozer.

Model 551 forklift.

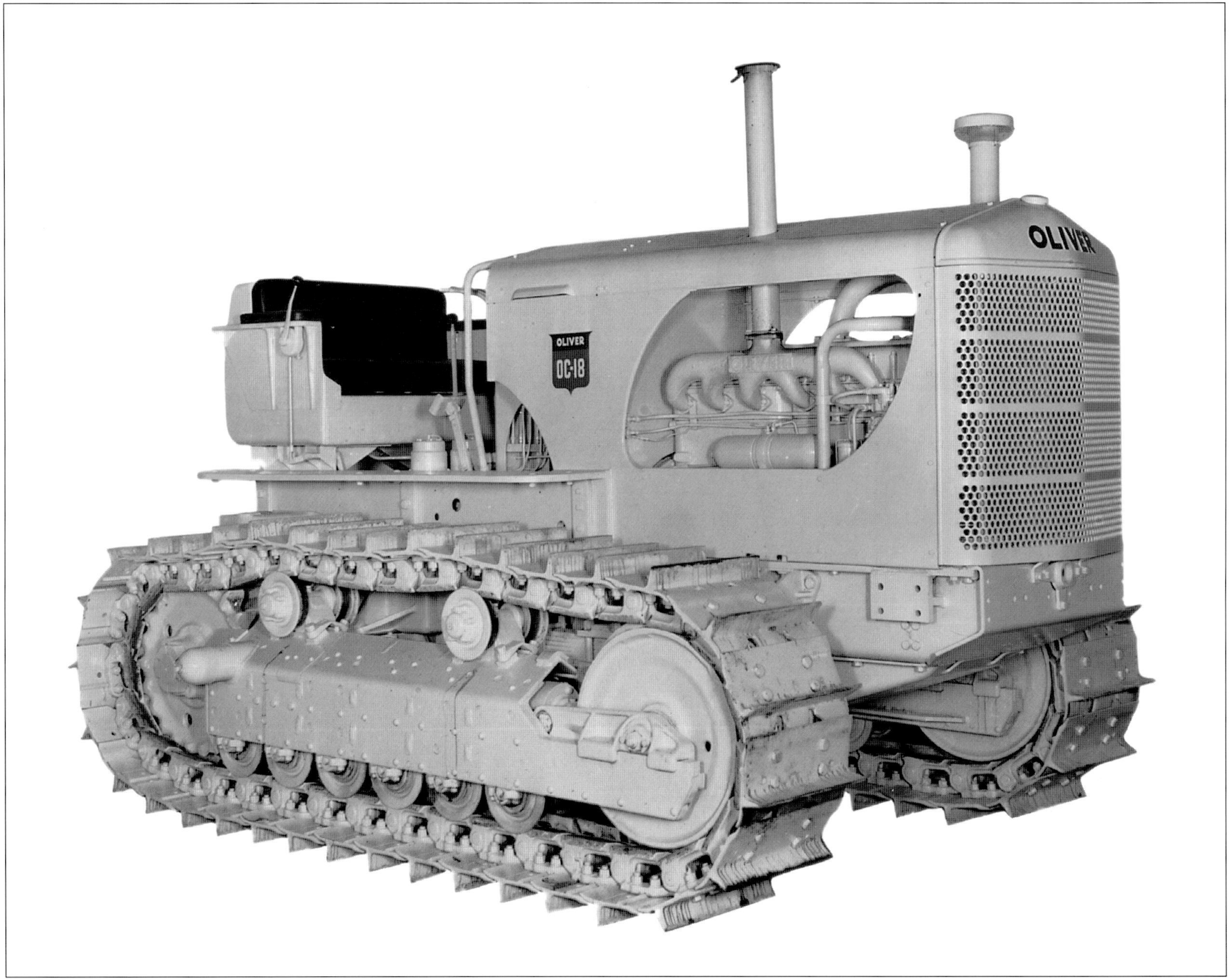

Two views of a Model OC-18 crawler tractor.

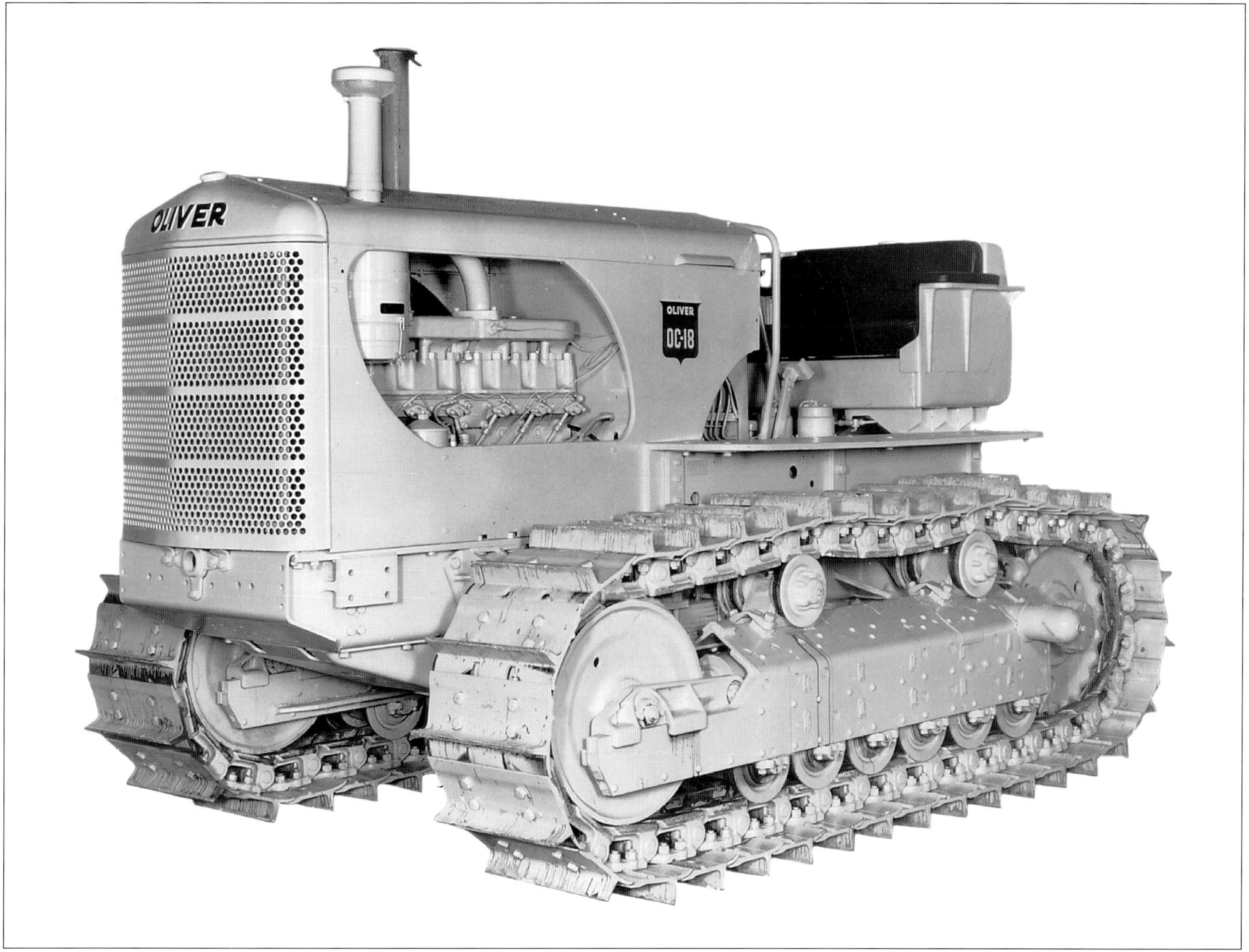
OLIVER
OLIVER
OC-18

Twin Oliver 880 industrial tractor and sheep-foot roller.

OLIVER
OLIVER
OLIVER
880

Chicago exhibit of Oliver industrial tractors. 1957.

STUDIO ARTWORK

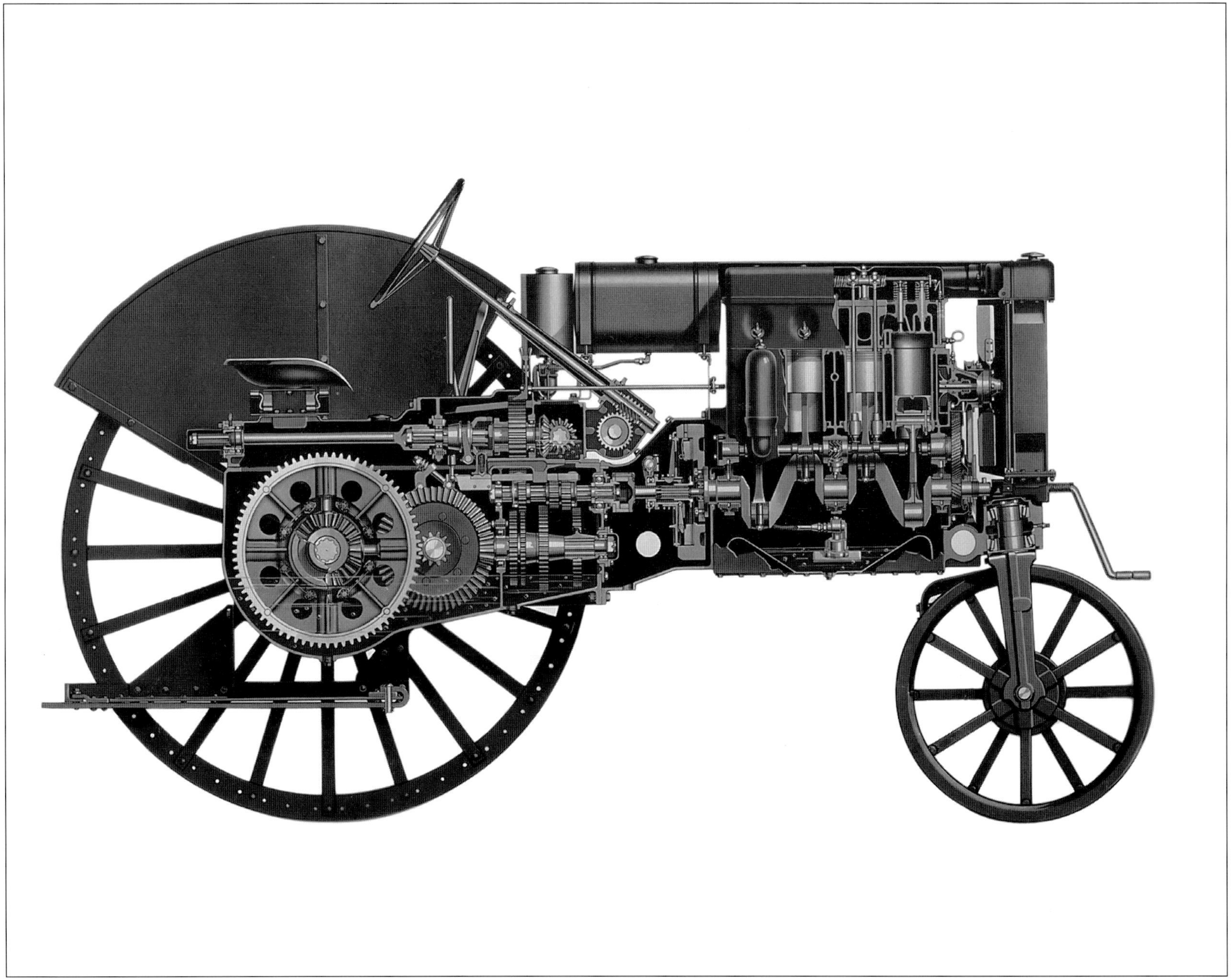

Cutaway of Oliver Hart-Parr Row Crop.

Oliver Hart-Parr 18-28.

Row Crop 60 with standard steel wheels.

ROW CROP
OLIVER
OLIVER
60

Row Crop 80.

Standard Model 80.

Left and right side views of an Orchard 77.

OLIVER
ORCHARD 77
77

Oliver Diesel Super 55 with sickle mower.

Cutaway of Super 77 gas tractor.

Super 88 Row Crop.

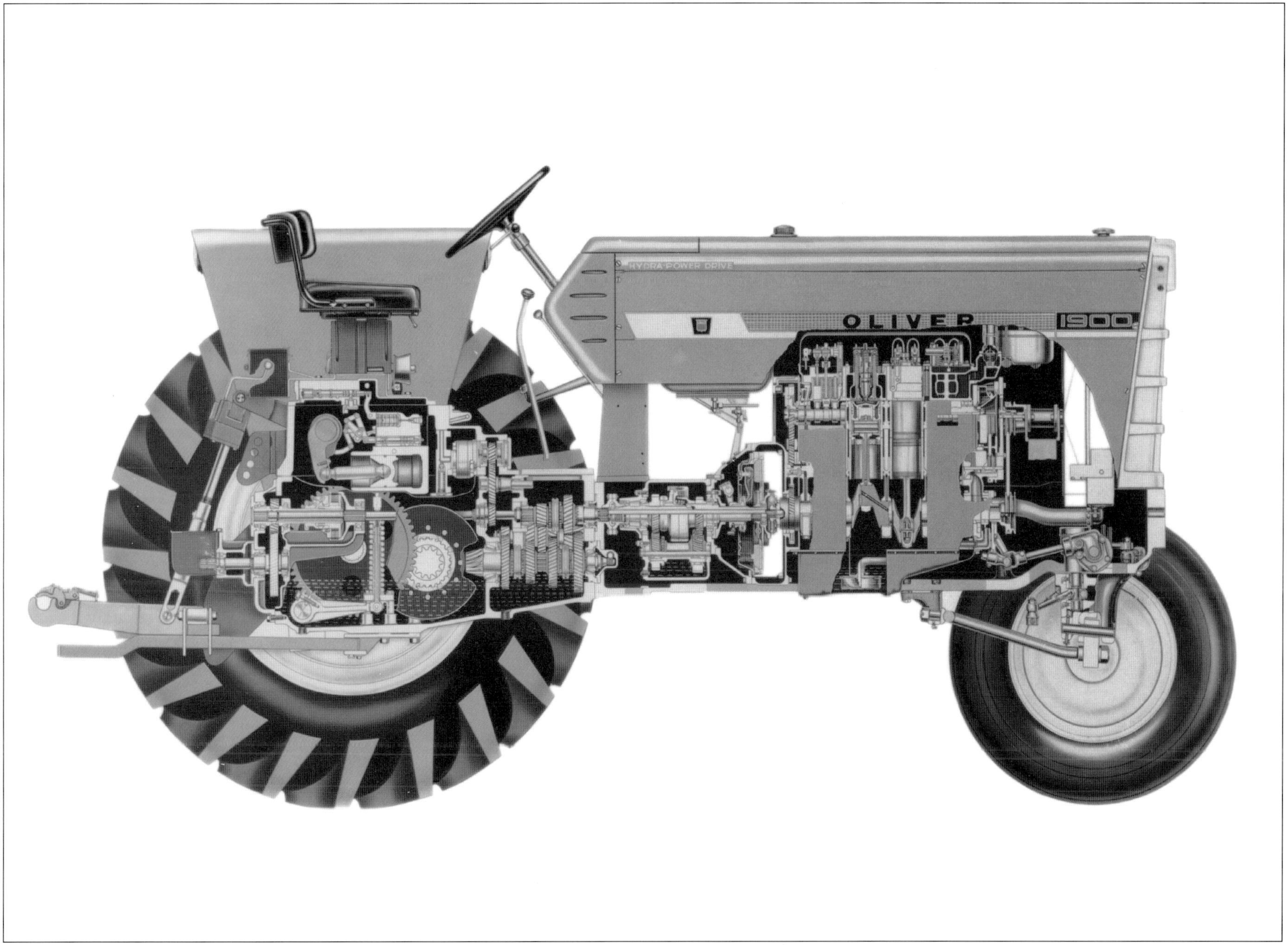

Oliver 1900 D cutaway. 1962.

Oliver 990 Speedhaul scraper.

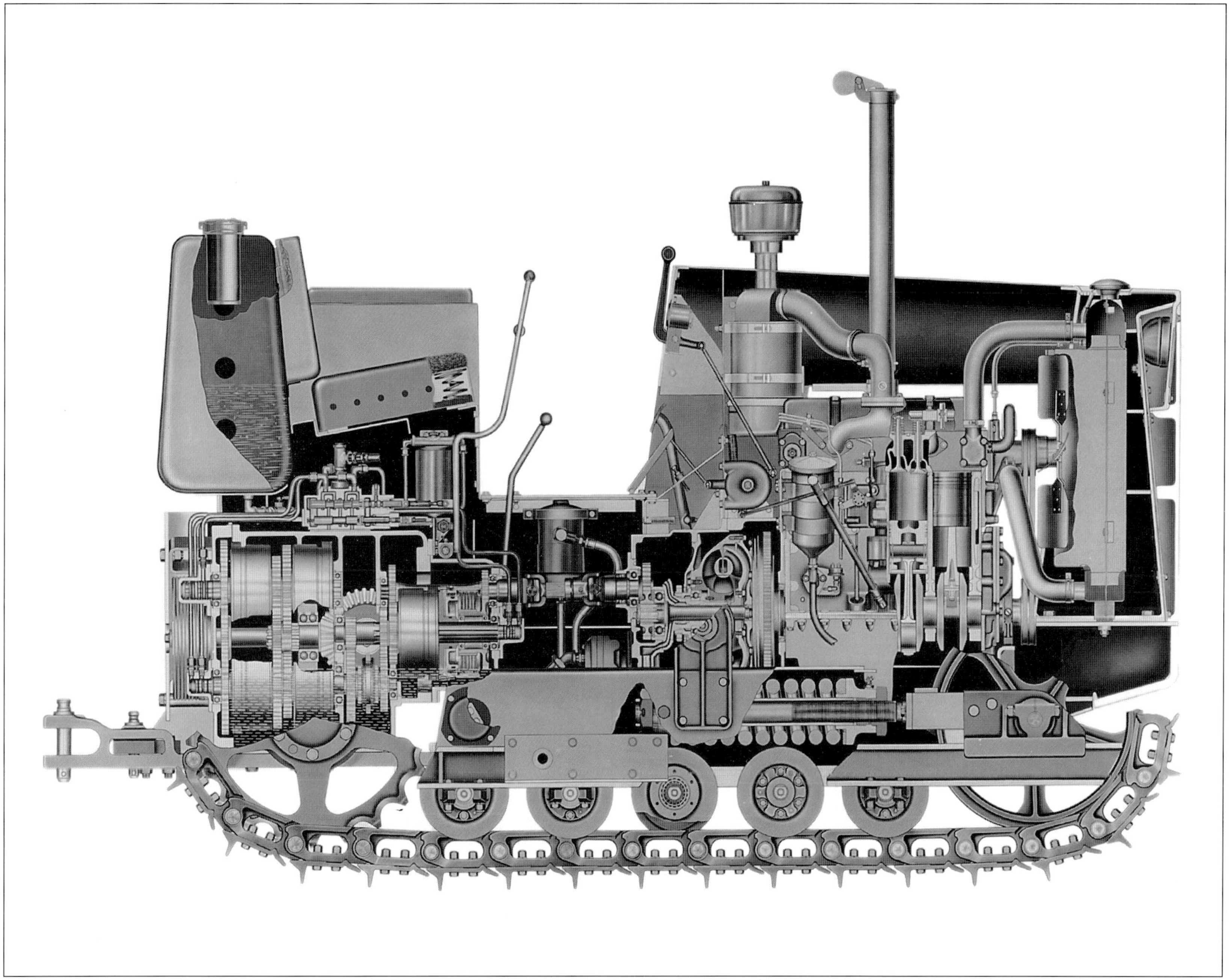

OC crawler cutaway.

Picture Credits

Floyd County Historical Society, Charles City Iowa: Pages 8, 12, 13, 14, 15, 16, 17, 18, 19, 20, 21, 22, 23, 24, 26, 27, 28, 29, 32, 33, 34, 35, 36, 37, 38, 40, 41, 42, 43, 44, 47, 52, 56, 57, 58, 59, 65, 68, 69, 71, 72, 90, 111, 113, 114, 115, 124, 125, 128, 129, 130, 131, 132, 133, 134, 135, 137, 138.

Higgins Collection, Shields Library, Special Collections, University of California, Davis: Pages 25, 30, 31, 39, 45, 46, 48, 49, 50, 51, 53, 54, 55, 73, 112, 116, 117, 118, 119, 122, 123, 136, 140.

Bakken Collection, Shields Library, Special Collections, University of California, Davis: Pages 60, 61, 62, 63, 64, 66, 67, 70, 74, 75, 76, 77, 78, 79, 80, 81, 82, 83, 84, 85, 86, 87, 88, 89, 91, 92, 93, 94, 95, 96, 97, 98, 99, 100, 101, 102, 103, 104, 105, 106, 107, 108, 109, 110, 120, 121, 126, 139, 141.

The Iconografix Photo Archive Series includes:

TRACTORS AND CONSTRUCTION EQUIPMENT

CASE TRACTORS 1912-1959 Photo Archive	ISBN 1-882256-32-8
CATERPILLAR MILITARY TRACTORS VOLUME 1 Photo Archive	ISBN 1-882256-16-6
CATERPILLAR MILITARY TRACTORS VOLUME 2 Photo Archive	ISBN 1-882256-17-4
CATERPILLAR SIXTY Photo Archive	ISBN 1-882256-05-0
CATERPILLAR THIRTY Photo Archive	ISBN 1-882256-04-2
FARMALL F-SERIES Photo Archive	ISBN 1-882256-02-6
FARMALL MODEL H Photo Archive	ISBN 1-882256-03-4
FARMALL MODEL M Photo Archive	ISBN 1-882256-15-8
FARMALL REGULAR Photo Archive	ISBN 1-882256-14-X
FORDSON 1917-1928 Photo Archive	ISBN 1-882256-33-6
HART-PARR Photo Archive	ISBN 1-882256-08-5
HOLT TRACTORS Photo Archive	ISBN 1-882256-10-7
JOHN DEERE MODEL A Photo Archive	ISBN 1-882256-12-3
JOHN DEERE MODEL B Photo Archive	ISBN 1-882256-01-8
JOHN DEERE MODEL D Photo Archive	ISBN 1-882256-00-X
JOHN DEERE 30 SERIES Photo Archive	ISBN 1-882256-13-1
MINNEAPOLIS-MOLINE U-SERIES Photo Archive	ISBN 1-882256-07-7
OLIVER TRACTORS Photo Archive	ISBN 1-882256-09-3
RUSSELL GRADERS Photo Archive	ISBN 1-882256-11-5
TWIN CITY TRACTOR Photo Archive	ISBN 1-882256-06-9

TRUCKS

DODGE TRUCKS 1929-1947 Photo Archive	ISBN 1-882256-36-0
DODGE TRUCKS 1948-1961 Photo Archive	ISBN 1-882256-37-9
MACK MODEL AB Photo Archive	ISBN 1-882256-18-2
MACK MODEL B 1953-66 Photo Archive	ISBN 1-882256-19-0
MACK MODEL B 1953-1966 VOLUME 2 Photo Archive	ISBN 1-882256-34-4
MACK EB, EC, ED, EE, EF, EG & DE 1936-1951 Photo Archive	ISBN 1-882256-29-8
MACK EH-EJ-EM-EQ-ER-ES 1936-1950 Photo Archive	ISBN 1-882256-39-5
MACK FC, FCSW & NW1936-1947 Photo Archive	ISBN 1-882256-28-X
MACK FG-FH-FJ-FK-FN-FP-FT-FW 1937-1950 Photo Archive	ISBN 1-882256-35-2
MACK LF-LH-LJ-LM-LT 1940-1956 Photo Archive	ISBN 1-882256-38-7
STUDEBAKER TRUCKS 1928-1940 Photo Archive	ISBN 1-882256-40-9
STUDEBAKER TRUCKS 1941-1964 Photo Archive	ISBN 1-882256-41-7

AUTOMOTIVE

AMERICAN SERVICE STATIONS 1935-1943 Photo Archive	ISBN 1-882256-27-1
IMPERIAL 1955-1963 Photo Archive	ISBN 1-882256-22-0
IMPERIAL 1964-1968 Photo Archive	ISBN 1-882256-23-9
LE MANS 1950: THE BRIGGS CUNNINGHAM CAMPAIGN Photo Archive	ISBN 1-882256-21-2
SEBRING 12-HOUR RACE 1970 Photo Archive	ISBN 1-882256-20-4
STUDEBAKER 1933-1942 Photo Archive	ISBN 1-882256-24-7
STUDEBAKER 1946-1958 Photo Archive	ISBN 1-882256-25-5

The Iconografix Photo Archive Series is available from direct mail specialty book dealers and bookstores throughout the world, or can be ordered from the publisher. For additional information or to add your name to our mailing list contact:

Iconografix
PO Box 609
Osceola, Wisconsin 54020 USA

Telephone: (715) 294-2792
(800) 289-3504 (USA and Canada)
Fax: (715) 294-3414